Ronald Stiff, whose models are shown on pages 14, 21, 36, 37, 39 and 40.

MODELLING FARM CARTS

John Vince

Shire Publications Ltd

CONTENTS

Copyright © 1977 by John Vince. First published 1977. Shire Modelmaker no. 1. ISBN 0 85263 405 6. All rights reserved. No part of this publication may be reproduced or transmitted in any form or by any means, electronic or mechanical, including photocopy, recording, or any information storage and retrieval system, without permission in writing from the publishers, Shire Publications Ltd, Cromwell House, Church Street, Princes Risborough, Aylesbury, Bucks, HP17 9AJ, U.K.

Printed in Great Britain by City Print (Milton Keynes) Ltd, Simpson Road, Bletchley, Milton Keynes, Bucks.

ACKNOWLEDGEMENTS

The author wishes to record his appreciation for the help and guidance of: K. Hansen; A. J. de la Mare; Richard Fowler; John Thompson; George Cusack; Len Hart, whose model is shown on page 21; and Ronald Stiff, whose models are shown on pages 1, 37, 40.

The cover photograph shows a model of a Kentish dung cart built by George Cusack using David Wray's plan.

A Tasker liquid manure cart in action. Very few vehicles of this type have been preserved.

A plank-sided tip cart turned out for a show.

INTRODUCTION

Throughout the British Isles two-wheeled horse-drawn farm carts of all kinds were once universally used. There were heavy carts with tipping bodies that could be used for practically any farming task and the light-bodied versions designed for work in the harvest field. One by one the old carts were pushed to the back of the cart shed and replaced by tractors with steel trailers.

The wheelwrights who made the carts used by our grandfathers did not work from a set of drawings. Using traditional tools, they worked with the help of a few wooden templates and the skills acquired from long years of experience, selecting the right timber for each particular part: elm for the hubs (naves) of the wheels,

oak for the spokes, and ash for the wheel rims and for the shafts.

Seldom does one find two carts exactly alike, although carts from the same workshop do share a family similarity. Local conditions played an important part in cart design. In the flat terrain of East Anglia carts were heavier than those built for use among the hilly lanes of Dorset and Devon. There are also numerous differences in detail.

In this book the main differences between tip carts and fixed-bodied carts are explained. Variations in cart design are included and many detailed photographs are provided to enable modellers to examine features which are not always immediately apparent from even the best scale drawings.

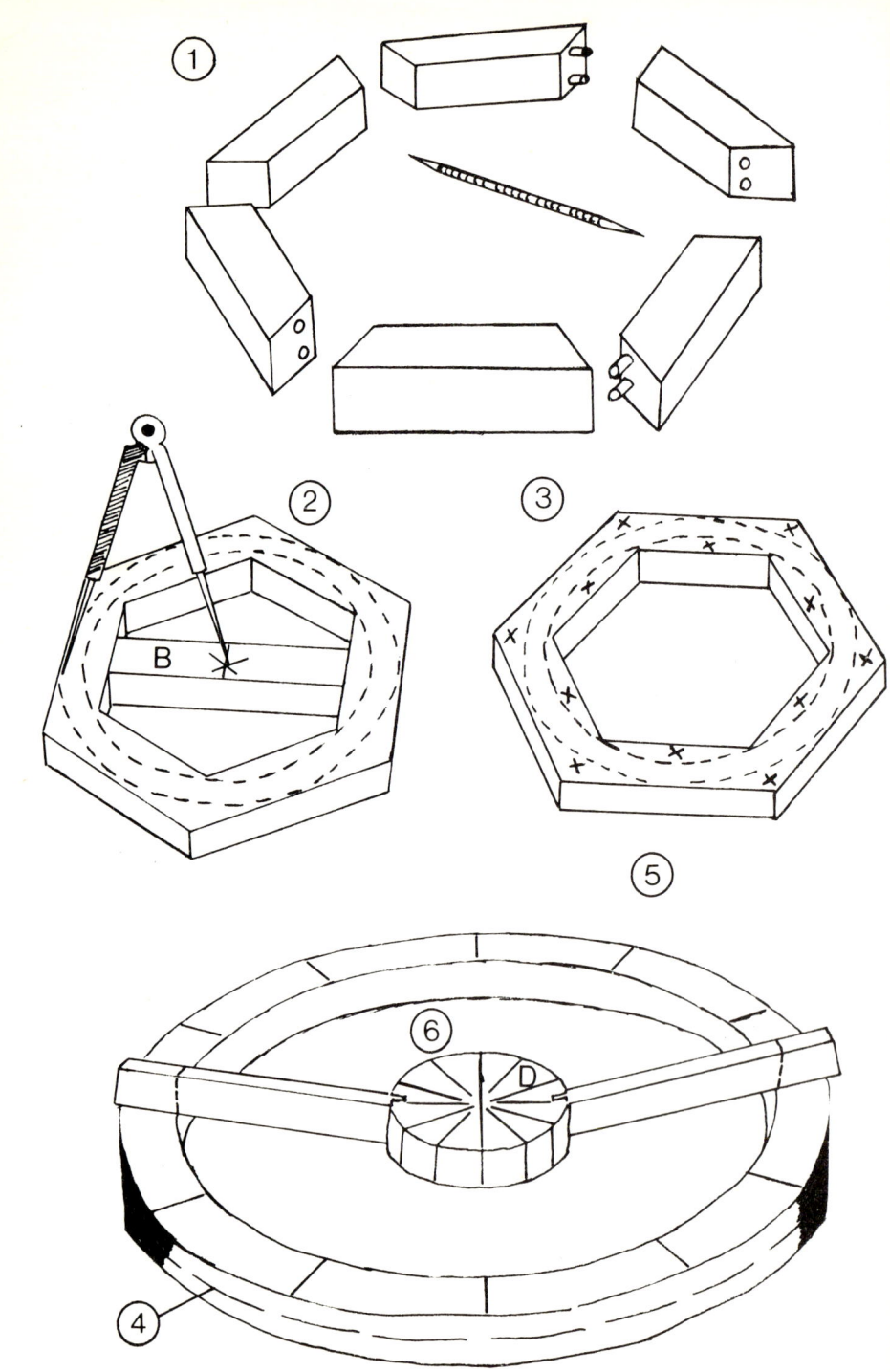

4

WHEEL CONSTRUCTION

'The traditional method of making a wheel was a time-consuming task. First, the wheel's hub (nave) was turned from well-seasoned elm. Then mortises were cut around its circumference to house the inner tenons of the spokes. When the spokes and outer rims (felloes) had been prepared the wheel was ready for assembly. Wheelwrights used templates to shape the curved felloes, but the construction of miniature wheels is easier if a different method is followed.

Modellers who admire the vehicles seen at exhibitions and in museums are sometimes discouraged from trying to build models of their own because the intricate construction of the wheel looks so complicated. There are several parts to a wheel and cart wheels almost always have a 'dish' to them. The method of wheel construction which is explained here in detail is well within the capabilities of anyone who can use a saw and a file. It was devised by Richard Fowler and enables the modeller to build wheels without using a lathe. On the illustration opposite the numbers in circles refer to the stages described below.

Stage 1. The construction of the wheel begins with the formation of the rim. A twelve-spoked wheel is made from a hexagonal rim, with each section (or felloe) cut at an angle of 60 degrees. Wheels with ten spokes are based on a pentagonal (five-sided) rim.

Each section of the rim is secured by two dowels. Cocktail sticks with a diameter of about 3/32 inch (2 mm) are very useful for this purpose. The dowels must be positioned carefully as shown, because later this thick rim is divided to make two wheels.

Stage 2. The pieces of the rim must now be glued together. During this operation the hexagon wheel should be securely cramped to a heavy baseboard. A block B is cut to fit firmly between opposite sides of the hexagon. Diagonals are then marked from opposite corners to find the centre of the wheel. With a pair of compasses the outer and inner edges of the wheel can be marked out.

Stage 3. When the block B is removed the unwanted portions of the rim (marked with an x) may be removed, using a saw and a file.

Stage 4. After a broad rim has been formed it can be divided into two equal parts by using a mortise gauge. When the line has been scribed around the circumference the wheel can be sawn into two. The sawn surfaces should be smoothed by being rubbed carefully on a flat sheet of fine glasspaper.

Stage 5. The position of the spokes can now be marked around the rim. There are two spokes to each felloe. Each hub is made up from four wooden sections. In this diagram the spoke disc D is shown divided into twelve spoke positions.

Stage 6. The spokes are prepared and a tenon is formed at one end. Always cut spokes with at least a quarter of an inch (7 mm) extra in length. The spoke disc can then have all the mortises filed around its edge with a fine file. Before starting to file mark the shoulder of each mortise by making a small cut with an Exacto knife saw. The most useful file for the purpose is one which has one smooth edge. A coarse grade of file is best as it will not become so easily clogged. This drawing shows two spokes in position.

Stage 7. A cart wheel is not simply a flat disc. The spokes of a wheel are set at an angle – called the dish. The angle of dish can be fixed by gluing four key spokes to the spoke disc and placing the spoke disc at the centre of the wheel so that the spokes rest on the inner edge of the rim and are set at an angle. When the spokes are firmly set mark out their true lengths as the diagram shows. Cut the spokes to length to make a push fit into the rim.

Stage 8. Use a clamp to hold the spoke disc firmly and drill a hole through the rim and into each spoke (see diagram below). Place a dowel in each hole to secure the spoke. When all the holes have been drilled glue the spokes and rim together. Trim off the end of each dowel.

Stage 9. Fit the remaining spokes in the manner described in stages 7 and 8. When a spoke is fitted fix its opposite member as the diagram shows.

Stage 10. When the main part of the wheel has been constructed the other sections of the hub can be attached.

These are shown in the diagram. A is a plain spacer piece. B and C are the outside sections of the nave. Their curved profiles can be prepared in the following manner.

Fix an electric drill in an appropriate bench attachment supplied by the makers. These attachments enable the drill to be held horizontally on the workbench. The dowel is drilled through the centre and attached to the drill chuck with a wire nail or bolt. With the electric drill in its clamp the rotating section of dowel can be filed into shape and then finished with sandpaper.

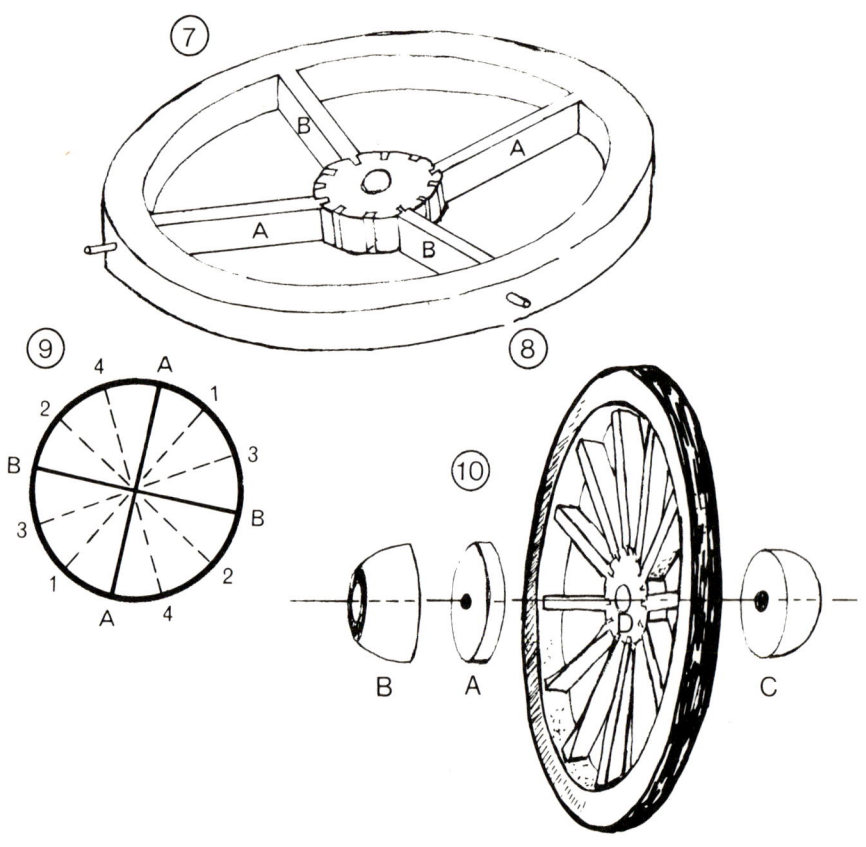

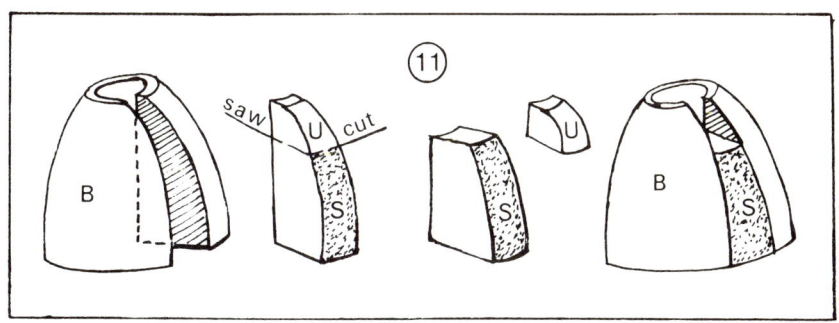

Stage 11. After it has been removed from the drill the internal diameter of section B can be enlarged. This may be done with a drill of suitable size, or the required diameter can be marked on the face and then formed with a round file. The size of this aperture will depend on the vehicle being constructed. There is no 'exact' size which is universally correct. A slice S can now be removed from section B as shown. This slice must be cut into two parts. Part S can now be glued back into place. Part U (the stopper) must be kept and fixed on to the wheel with a clasp as shown above.

Stage 12. The hub is now almost ready for assembly. Sections A and C may now be glued to D. Cut a length of brass tube equal to the thickness of sections A, C and D, plus about half the thickness of B (see diagram above). This will be the 'box' in which the axle will run. The tube can be glued into position with Araldite. If necessary the box can be trued up by using

the pointed ends of the cocktail sticks as wedges before the glue is applied.

Once the box is in the correct position section B can be glued into place and the hub is then complete.

FINISHING THE WHEEL

The final task is to fit the stopper (U above), which is secured by a stopper clasp. Stopper clasps vary in shape and the modeller will find it useful to collect as many illustrations of such details as possible. It is helpful to keep a notebook so that rough sketches of details of this kind can be recorded for future reference.

The stopper clasp is made from a flat sheet of tin or brass (see below). Cut a strip of the appropriate width and use a scriber to mark the profile. Drill the hole for the pin P before the profile is shaped. A $\frac{1}{16}$ inch (2 mm) drill is suitable for this purpose. To file the stopper to the required shape place it in a small vice or hand vice,

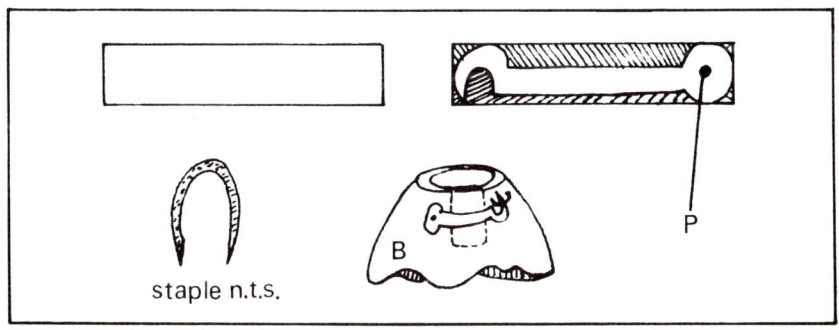

staple n.t.s.

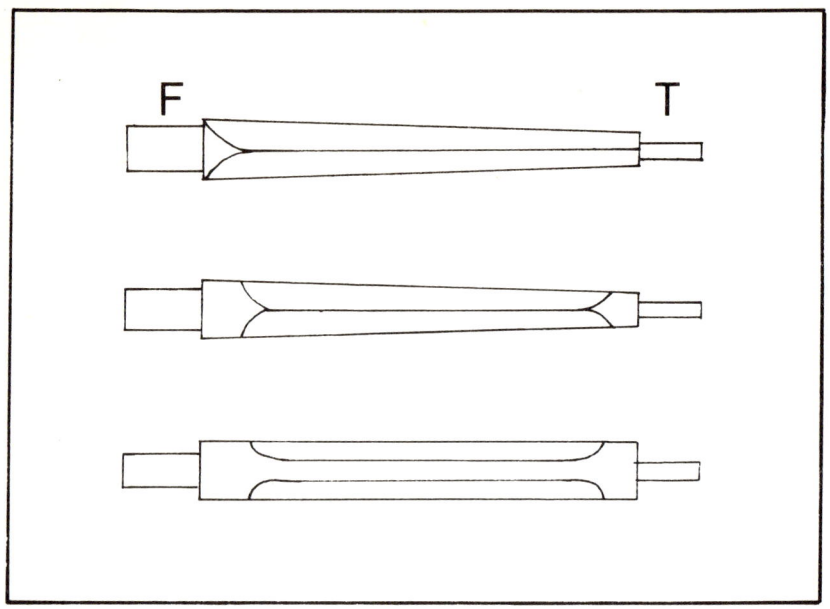

or use a pair of flat-nosed pliers.

The staple can be made with a panel pin bent into a U shape. File the ends of the staple into a point. To make it easier to attach the pin and staple, clearance holes of about $\frac{3}{64}$ inch (4 mm) can be made in the hub before they are attached. Care is needed for this operation as it is very easy to split the grain apart if insufficient clearance is allowed or too much force is used. Experience alone will enable the modeller to achieve the delicate touch this operation demands. It is advisable to fix the stopper clasp in position after the wheel has been painted.

The spokes of a wheel are originally made from a rectangular section. After the wheel has been assembled the spokes may be finished with the help of a fine knife. Tools can be made for this purpose from bits of hacksaw blade. The Swann-Morton Craft Tool fitted with a number 3 blade is very suitable for this purpose – see list of suppliers of tools and materials on page 46.

Above: Front view of spokes showing alternative methods of chamfering (F foot, T tongue).

TYRING THE WHEEL

Fixing the tyres is an important operation. There are two types of tyre. Strake tyres are made up of separate sections which are nailed on to the felloes. Hoop tyres present the modeller with a problem as it may not always be easy or convenient to weld, braze or solder a hoop tyre together on the kitchen table. The simplest method of fixing a tyre to a rim is to cut a strip of mild steel to an exact length so that its square-cut ends make a butt joint. Each joint can then be secured by a countersunk pin, and then the pin heads should be filed flat. This method is perhaps adequate but not very satisfactory in

8

Below: Wheels from a Cornish tip cart. The chamfering almost meets in the middle of each spoke.

Above: An iron nave with slightly chamfered spokes.

its appearance, which cannot be disguised.

The silver-solder method shown here is probably the best way of tyring as it requires very little equipment. Silver solder is expensive but each tyre requires so little that the cost is not more than a few pence. This contrasts dramatically with the considerable expense of an electric welding or brazing unit.

SILVER SOLDERING

When the metal tyre has been cut to a precise length its two ends can be soldered together. A tyre that has been pushed on to a wooden rim has to be in tension. To prevent the soldered ends from separating under pressure silver solder must be used. This, and the necessary flux, can be obtained from K. R. Whiston (see page 47).

The stages of making the soldered joint are as follows:

1. On the inside and outside of the butt joint file a V-shaped channel. This will help the solder to find its way into the joint.
2. Make sure the joint is clean and free from grease or other foreign matter; then add a globule of saliva to the joint.
3. Add flux to the saliva.
4. Cut a length of solder equal to the width of the tyre and place in position in the V-shaped groove.
5. Slowly heat the tyre with a flame until the solder melts. Its melting point is about 760 degrees Centigrade.
6. When the joint has cooled, clean off any surplus solder with a file. If you have no previous experience of using solder it is worth practising making some joints before you make up real tyres.
7. The completed tyre is now ready to be pushed on to the wheel. A slight bevel can be made on the outer edge of the wooden rim to give the tyre a 'start' if necessary.

Silver soldering a metal tyre. Below left: adding flux to the joint. Below right: placing a length of solder in the groove. Bottom left: heating the joint. Bottom right: the finished tyre.

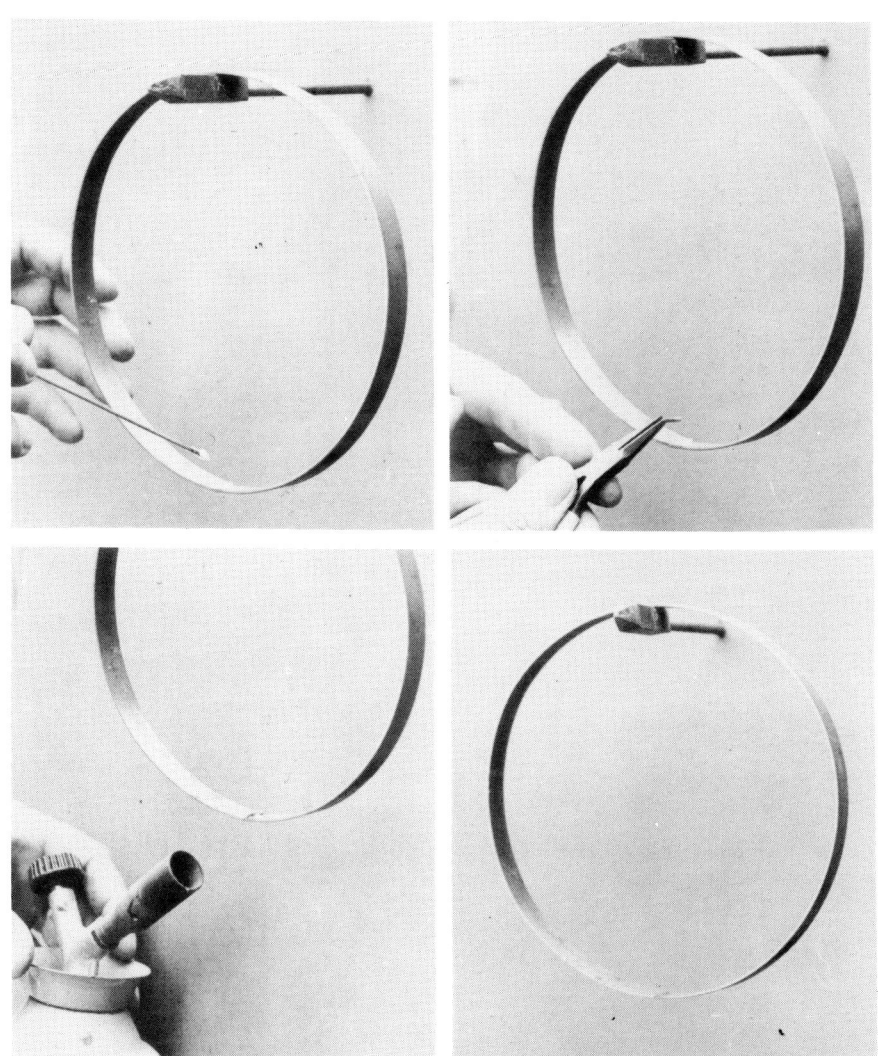

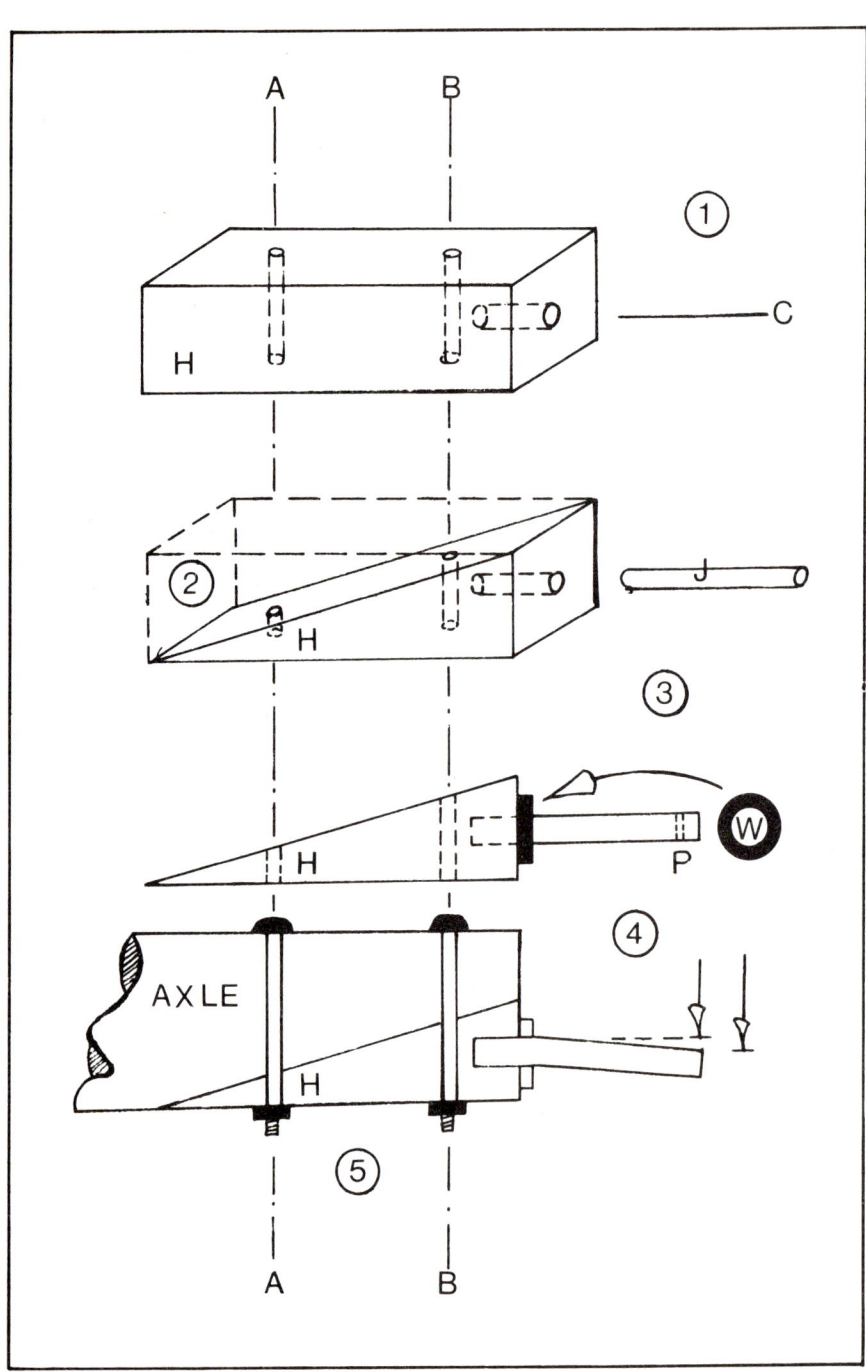

THE AXLE ARM

Two types of iron axle are found on carts. Some carts have a long forged axle which is attached to the axletree with long bolts. An example of this type is used on the Cornish cart shown on page 24. Through axles were also employed on spring carts.

The alternative to the through axle was the shorter axle arm. This arm was also fixed to the wooden axletree. John Thompson's plan of a Cotswold cart (see illustration on page 33) has axle arms of this kind.

A model axle arm can be made in two parts, i.e. from a $\frac{1}{4}$ inch (7 mm) square section of bar and a round $\frac{3}{16}$ inch (5 mm) diameter rod. Both these items can be obtained from the appropriate suppliers listed on page 46.

To prepare an axle arm, cut a length of bar to the required length. Then cut a section of rod long enough to allow for the portion which fits into the bar. When these items are ready follow this sequence as illustrated in the diagram opposite.

1. Mark and drill two holes A and B in the block H. These holes will eventually house the long bolts that hold the arm to the axletree. When this has been done drill hole C into the end of the block. The journal J will be soldered into this hole.

2. Once the three holes have been made the block H can be filed down to the triangular shape shown.

3. Solder the journal into position. The washer W should then be soldered on to the inner end of the journal. This strengthens the arm and provides a face for the inside of the nave to bear upon.

4. When the arm has been assembled the hole for the lynch pin P should be drilled in the journal. On a full-sized vehicle the journal is not parallel to the ground; it slopes slightly downward. This dip in the alignment of the journal can be made by placing each arm in a vice and bending each gently downwards. Make sure that all the axle arms for a particular model are given the same degree of dip. A journal an inch (25 mm) in length should have a dip of about $\frac{1}{8}$ inch (3 mm).

5. After the arms have been made, the wooden axle can be prepared and the sloping grooves to take the arms cut with a fine chisel. The bolt holes in the wooden axle can then be marked and drilled.

The cart wheels below are assembled on a through axle which allows us to see the downward dip of the journals. A is the centre line of the axle. B is the centre line of the naves.

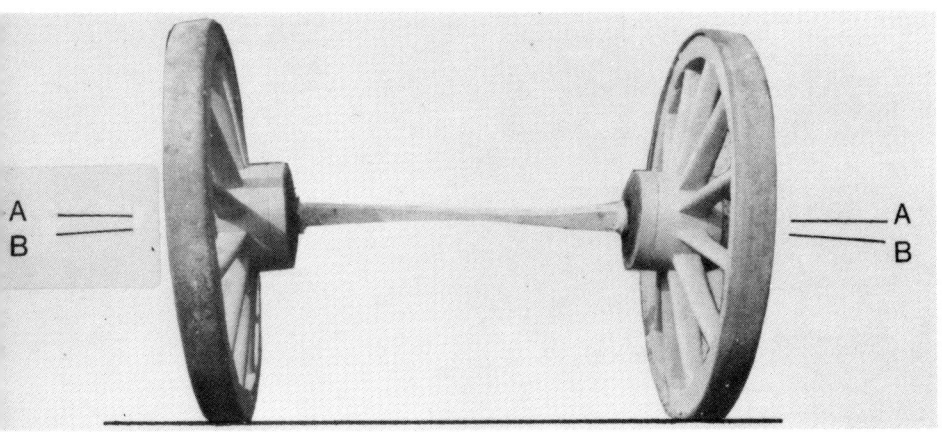

A Wiltshire tip cart which has much simpler lines than the Kent cart shown on page 15. Three plain strouters support the sides and there are no body standards. Notice how the shutter linking the shafts lacks the strength provided by a mortise and has iron braces instead.

TIP CARTS

The typical medieval cart seems to have been rather crude. It was heavy and probably cumbersome as it usually required more than one horse to pull it when it was loaded. There are disadvantages in working a fixed-bodied cart. The quickest way to discharge a load is to tip the cart backwards. To do this with a fixed-bodied cart the horse has to be released from the shafts.

Tip-up bodies were probably introduced in the 1830s but no particular person is credited with the original idea. The method of tipping the cart body may have been borrowed from Flanders in the years before Victoria's accession.

Cart bodies are either long-boarded or cross-boarded. The floorboards of dung carts ran from front to back – this is called long-boarding – because the farmworker's shovel could be more easily used where there were no uneven joints crossing the cart floor to catch the blade of the shovel and cause it to stop abruptly and jar the worker's hands.

Long boards have to be supported by cross members (shutters) spaced between the front and rear shutlocks. On some carts a single intermediate shutter is used but most designs had at least two. Another special feature of dung carts is the uneven width of the cart body: the rear end is usually about two inches (50 mm) wider than the front. This difference allowed a heavy load to slide out unimpeded when the cart was tipped up.

A tip cart would be used for a variety of loads. Some designs had removable boards that deepened the sides and headboard in order to improve the capacity of the body. They often had ladders that could be added to front and rear for harvest work.

BODY CONSTRUCTION: LONG-BOARDING

In the sequence of pictures and diagrams which follows the main stages of body construction of a long-

14

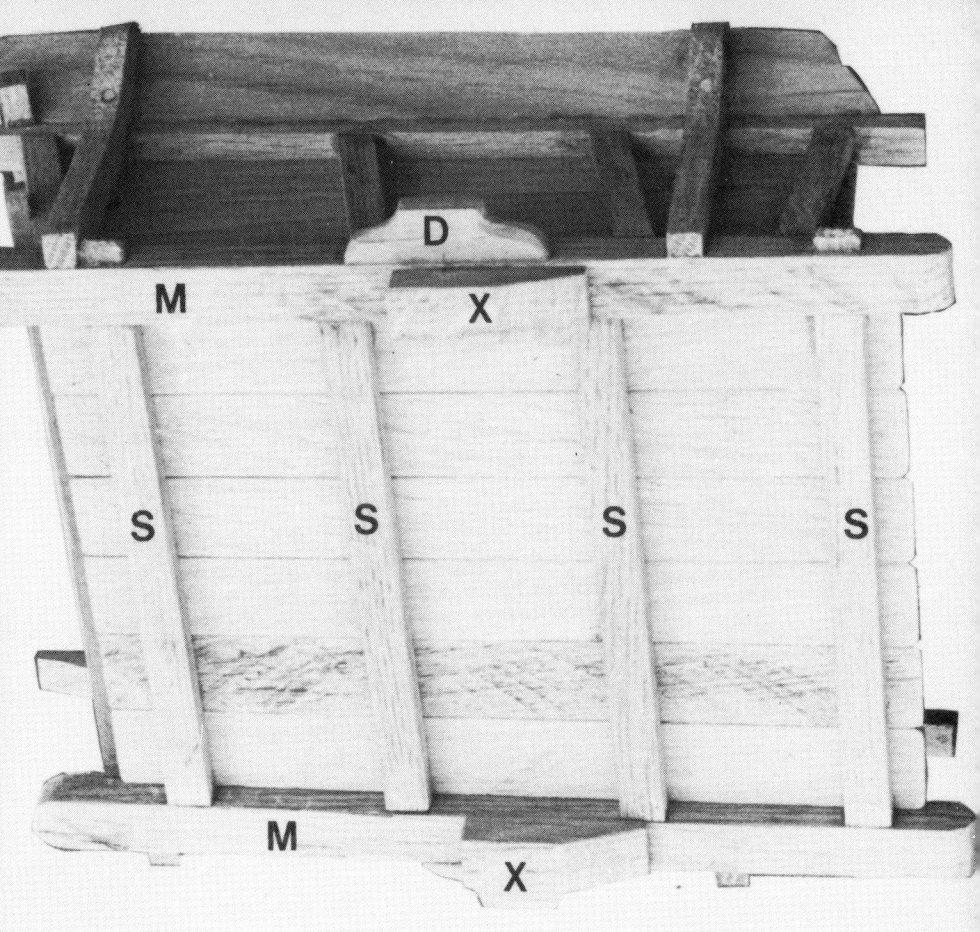

boarded Kentish tip cart are outlined.

The sequence does not necessarily represent the only method that may be used. All modellers will evolve their own methods and there is no single correct way. Part of the interest of building comes from the technical problems which occur, for instance when you suddenly discover that you have forgotten to bore a hole in a section which has, after assembly, been obscured by other parts of the vehicle. Many of John Thompson's plans contain clear instructions concerning the stages at which particular holes should be made; such detailed direction is always very helpful to the modeller.

The method described here has been evolved by George Cusack, using one of David Wray's plans.

The above view of the Kentish tip cart allows us to see clearly how the bed of the cart body is constructed. The main sides M are joined by four shutters S. On this rectangular framework the sides and front end of the body are assembled. Two blocks X are attached to the underside of the main sides. These blocks will eventually be fixed to the axle. On the outside of the members M, where the body crosses the line of the axle, there is a small block D. This is the dirt rave, intended to prevent dirt from entering the narrow space between the axle and the inside of the nave. The long floorboards are arranged so that they rest upon the shutters and between the main sides.

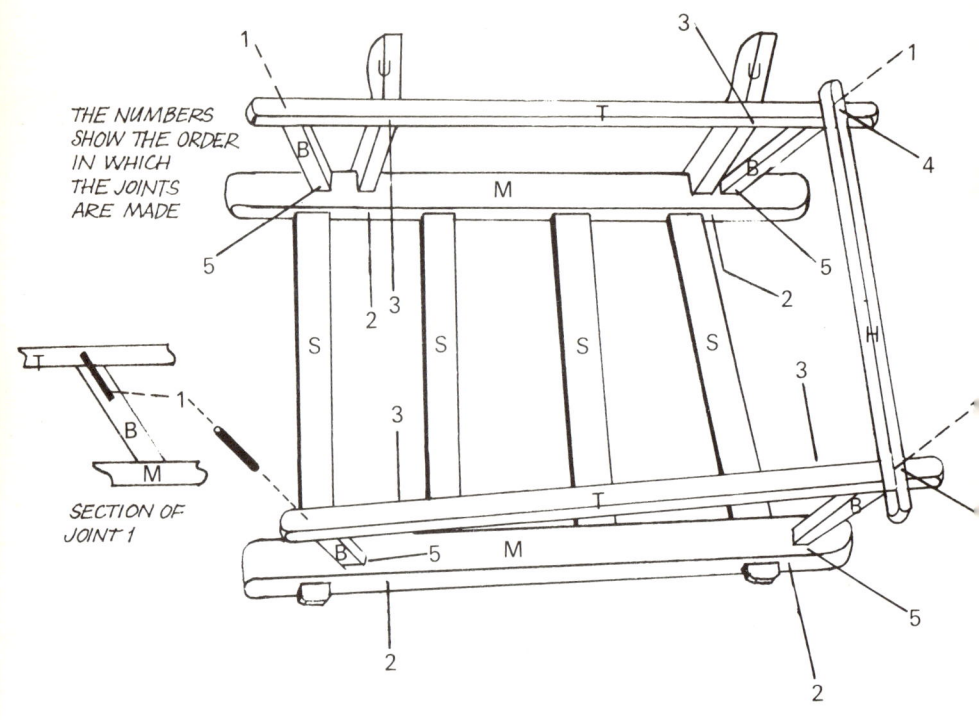

THE NUMBERS SHOW THE ORDER IN WHICH THE JOINTS ARE MADE

SECTION OF JOINT 1

ASSEMBLY SEQUENCE OF CART BODY

Stage 1. When the top rails T, the body standards B, and the strouters U have been prepared, the headrail H should also be cut to size. Drill the bolt holes in each strouter.

Stage 2. Assemble the side top rails by drilling appropriate holes through the rails and into the front and rear standards. A dowel (cocktail stick) can then be inserted and glued to make each joint. Build each side in this manner.

Stage 3. When the body sides have dried, the main framework of the body can be assembled. Drill the bolt holes in the main sides M which will secure the lower end of each strouter U. Position each strouter, and bolt them to the main sides M.

Stage 4. With the strouters in place the framework of each side can be positioned and joined together by the headrail. Glue the lower end of each standard to make a butt joint. Bolt the upper end of each strouter to the top rails. While the glue is setting keep the sides and headrail in place with appropriate blocks and small clamps. Take care not to overtighten them as this can cause distortion or even damage to the frame.

Stage 5. After the framework has dried the shaping of the standards, strouters and other rails can be carried out. For this purpose use a fine knife, such as the Swann-Morton craft knife with a number 3 blade.

Stage 6. When the shaving has been completed fit and glue the floorboards, sideboards, frontboards and standards Y, in that order. Fix the curved headboard last of all. Here is a view of the body from above. The numbers indicate the order in which the boards are fitted.

Below: The tailboard. Its brace A could be glued in place after the two pins shown in the photograph on p.19 have been fixed. This will allow it to be positioned to make a good fit against the eyebolt.

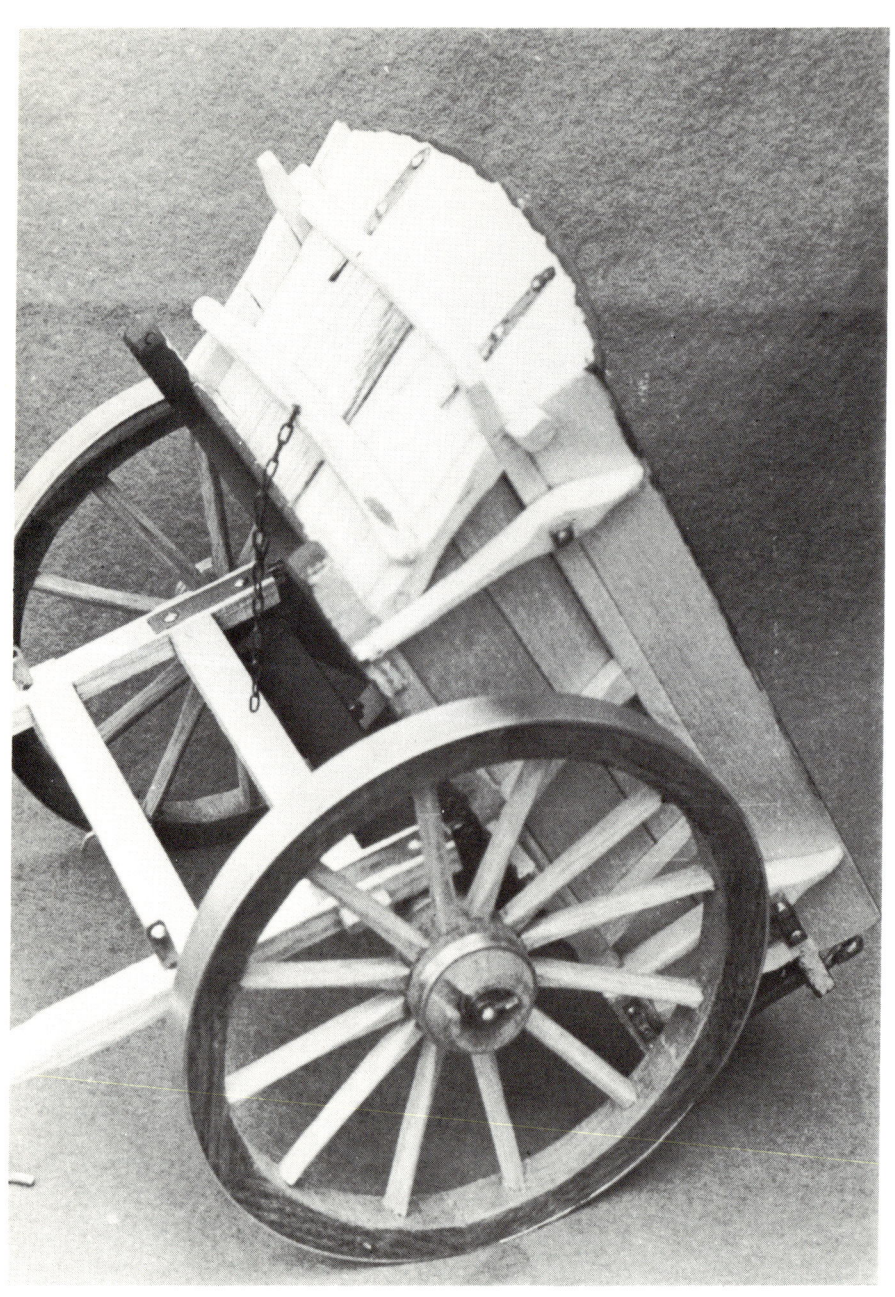

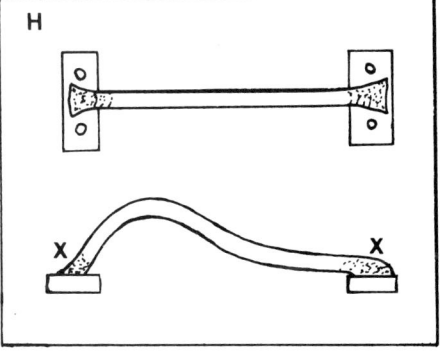

Above: A rear view of the completed cart. The hindstaff H is made from a round rod (see right) which has been flattened at each end and soldered at points X to two flat metal plates.

Left: A view of the cart showing how the shafts are hinged at the end by means of two interlocked staples. Notice too the chain used to restrict the angle of tip.

Above: A front view of the cart showing how the tipped body could be held with the chain attached to the front shutter of the shafts. The strap-stick is resting between them.

BODY CONSTRUCTION: CROSS-BOARDING

The second type of tip cart has cross-boards, which are parallel to the axle. This type of construction requires a different arrangement of supporting timbers. Instead of the shutters used on the dung cart the cross-boarded vehicle has longitudinal summers which run parallel to the sides. There is a shutlock at front and rear.

The fine model opposite, by Len Hart, has a cross-boarded bed. It was built from John Thompson's plan. The original vehicle is at the Weald and Downland Open Air Museum, Single-ton, West Sussex.

SHAFTS

The shafts of a cart have to bear many stresses. They could only be braced with cross members (shutters) at the rear end (see opposite). Various kinds of metal braces were added to the shafts to strengthen them. During a cart's working life a shaft often became damaged or broken, and a new end was then made. This was attached to the original shaft by means of a scarf joint. When a scarf joint was formed, a metal strap was placed above and below the shaft. Bolts were used to fix the two straps together and the wooden joint was then held in a vice-like grip.

TIPPING DEVICES

There are several different methods of controlling tipping. Some carts simply have a chain to control the degree of tip. The most sophisticated method was the screw-jack, which appeared in several trade catalogues towards the end of the last century. It must have been very slow in operation and this may explain why it is seldom seen. The other methods illustrated are: the bar, the strap stick, the lever, the sword and lever, the sword and sliding ring, the sword and peg.

Below: Bars were probably the first and simplest form of lock. On this example the shafts are set between the main sides of the cart. When the bar is in position it prevents the main sides from moving upwards. The weight of the body pushing upwards forces the bar against the staples. To remove the bar the retaining pin, on its chain on the viewer's right, must be withdrawn. The tapered bar can then slide outwards towards the vehicle's near side, if no spoke obstructs its passage.

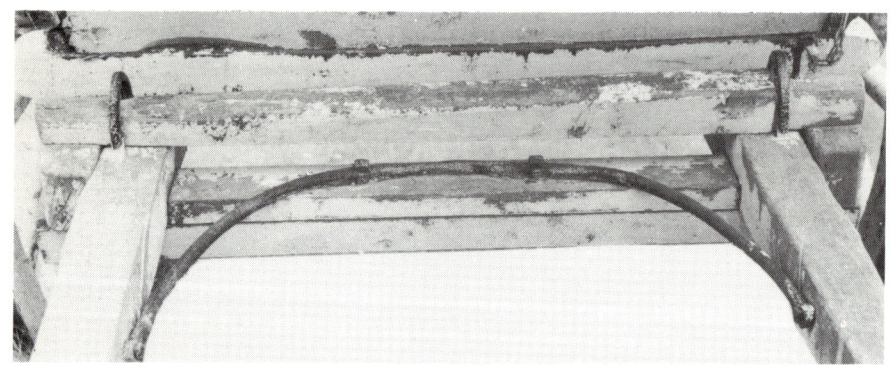

Below: The strap stick is a variation of the bar method and works on a similar principle. The metal ends of the strap stick are offset so that gravity will keep it in position between the extended main sides of the body. The cart sides on this vehicle are resting on the shafts. To remove the strap stick it is pulled forwards and upwards. The ends can then be disengaged from the eyes by moving the bar sideways until one end is released.

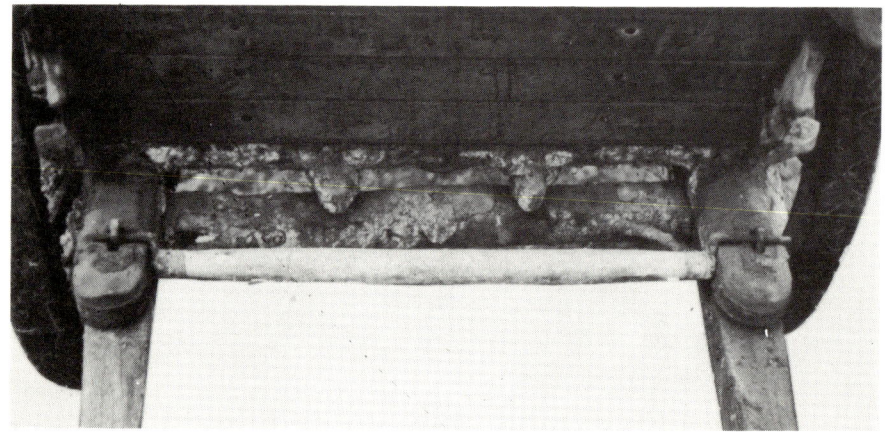

The principle of the lever could be applied to tipping gear. In this example the lever is pivoted at the centre line of the cart. The main sides of the cart are positioned outside the shafts. When the lever is released from the two opposed clips positioned on the shafts, the body is unlocked and able to move upwards and backwards. A pin on a chain is used to secure the lever so that it cannot be jolted out of its clip. This photograph shows the near side of the lever in the released position.

This view shows the lever principle combined with a vertical 'sword', which allows several different tipping positions. The lever is secured to the extended main side of the body by a horizontal clip. A strong spring S attached to the front shutlock keeps a constant pressure on the lever and this holds the pin end P in any hole selected on the sword stick T. A locking device 3 with a pin on a chain is used to prevent the lever jerking itself free from its clip when the cart is in motion.

Above: Another version of the sword stick was combined with a sliding ring which could be controlled by a spring or the more usual pegs.

Left: A very simple sword, or tip stick, was widely used in the south-western counties. The Cornish dung cart shown here has a curved block, attached between the two shaft shutters, which houses the lower end of the tip stick. When a suitable tipping position has been selected a peg or pin is placed through the hole in the tip stick. A second, identical pin is used to prevent the first pin from falling out of its position in the sword.

The capacity of this cart is increased by the addition of side and front boards. A scale plan of this vehicle is available from the author (see page 46).

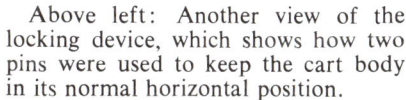

Above left: Another view of the locking device, which shows how two pins were used to keep the cart body in its normal horizontal position.

Above right: Not all tipping devices were positioned centrally. This sword stick is fixed to the nearside shaft. Two pins are used to keep the body in the required tipping position.

Below: One method of pivoting the tipping cart body was to rest it upon the shafts. A concave shape was made in each shaft. The intermediate shutter on the body had its ends made to fit this depression and in this very simple manner a 'hinge' was formed.

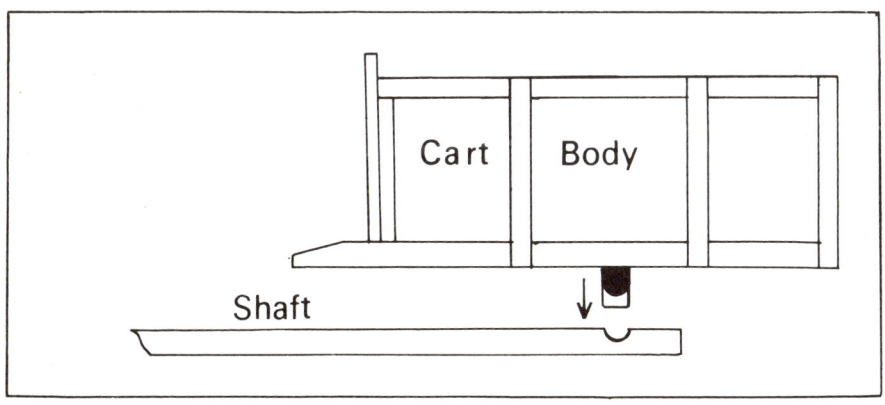

Cart Body

Shaft

An alternative form of hinging the body was to place the shafts inside the main sides and extend them beyond the axle. As this example shows, a rod could be used to link the main sides to the ends of the shafts to form a pivot.

If a cart body was tipped when loaded and struck the ground the tail end could suffer a severe jolt. To protect the body the main sides, the heaviest members, were often extended. This example shows how ironwork was used to add strength to the bumper. The two summers have also been made to protrude and, without looking inside, we know that this cart has crossboards.

Some carts had a prop stick attached below the rear shutlock to prevent the body being tipped up by accident. When it was not in use the prop stick was fixed to the axle with a short chain. Refinements of this kind are fairly easy for the modeller to add.

LONG CARTS

Long carts were mainly intended to be used in the harvest field, where bulky but light loads had to be carried over ground that was often uneven, though they did have other uses. On steep hillsides it is very difficult to turn a wagon but a two-wheeled cart can be handled more easily.

The long-bodied cart seems to have been invented, or introduced into the British Isles, in the late 1820s. The idea was subsequently copied in many of the hilly regions, where quite distinct versions were evolved.

One of the earliest long carts was built by a farmer named Henry Hannan, who used carts of this kind at his farm at Burcot, Oxfordshire, in 1828. His version was an adaption of carts then in use in Cumberland. Robert Robertson, a farmworker in

Fife, is credited with a similar design in 1832.

Some interesting names are used for these attractive designs. In Devon they are known as ladder carts, curry carts, and kerry carts. In the Cotswolds they are called Scotch carts, but this name was also used in East Anglia for heavy tip carts. To Cornishmen they are wains and the hill farmers of Wales know them as gambos. A few rare examples seem to have been made to tip.

Another variation on the long cart is the wheel car, sledge, or slide car. Its low-slung narrow body enabled it to be used on hillsides too steep for a conventional cart. Vehicles of this kind were used in Scotland, Wales, on the Pennines and in Devon.

This Devon sledge has been repaired with modern sawn planks on its off side, but the original portion was made from timber which was carefully chosen for its shape.

In Wales slide cars usually had their wheels placed in the middle. There is an excellent drawing of an example from Radnorshire in James Arnold's *Farm Waggons and Carts*.

A very simple sledge from the hills of Wales.

When the sledge was given a pair of wheels it became a truckle cart. Its body is 7 feet (2.1 m) long. A plan of this vehicle appears in John Thompson's list.

This North Devon ladder cart (above) is a variation of the cart shown in the diagram on page 16. There is however an interesting distinction in the manner in which the body of this vehicle is constructed. The two summers and main sides rest upon the axle and they are bolted to the shutters which oversail on each side. At the sides however there are lighter 'false sides' in which the vertical spindles are housed. The shafts are attached inside the main sides, and at the front they are secured with U bolts. The front-side spindle is a long bolt with an eye at the top and this feature was probably intended to provide an anchor point for a front ladder. Inside the body, short planks placed behind the wheels helped to prevent the load spilling on to the axle.

Left: This detail of the hindstaff helps us to understand why the shutters oversailed the sides. If the shutters finished exactly at the main sides there would be nothing to provide a support for the buttress, with its interesting barley-sugar twist.

This ladder cart, now at the James Countryside Museum, Bicton Gardens, provides an interesting contrast with the other Devon example opposite. The branding shows that it was used about twenty miles away from the East Buckland cart. The Highampton version still has its four shutters oversailing the sides but it has quite a different appearance. At front and rear the ends of the shutters are linked with a spindle, which adds some strength to the sides and helps to protect the staples holding the ladders. The sides fill the space between the intermediate shutters and they are supported by forestaffs and hindstaffs like those shown on the previous example. At the base the spindles are housed in the false sides noted on the other example. The bed of this cart is raised well above the axle on a block.

Above: This harvest cart is very similar to the Cotswold harvest wain (Sheet 19) in John Thompson's series of plans. However it is a heavier looking vehicle with cross-boards; hence the two summers. Removable ladders on the sides may have been later additions to the original structure. The two staples fixed to the front of the nearest side ladder were clearly intended to be used to support harvest poles, which added stability to a high load. Two other staples which could have served the same purpose are situated close to the fore and hind shutlocks. A duplication of this kind suggests that the owner found the cart of more use with ladder sides and poles in slightly varying positions. From a careful study of details an observer can sometimes discover clues which show how a vehicle was modified dur-

ing its working days. Freda Derrick illustrated a cart of this type in her book *Country Craftsmen* (1945). The modeller who is interested enough to undertake his own research can collect a mass of information from printed sources.

Opposite: The top photograph shows the Cotswold harvest wain measured and drawn by John Thompson. There are fittings at front and rear to accommodate harvest ladders or poles. Below the broad, hooped sideboard there are three hooks which were provided to secure the ropes.

Below is a view of the long boards showing how they curve upwards from the centre towards the front and rear shutlocks. A dished body of this kind helped to lower the load's centre of gravity and increased its stability.

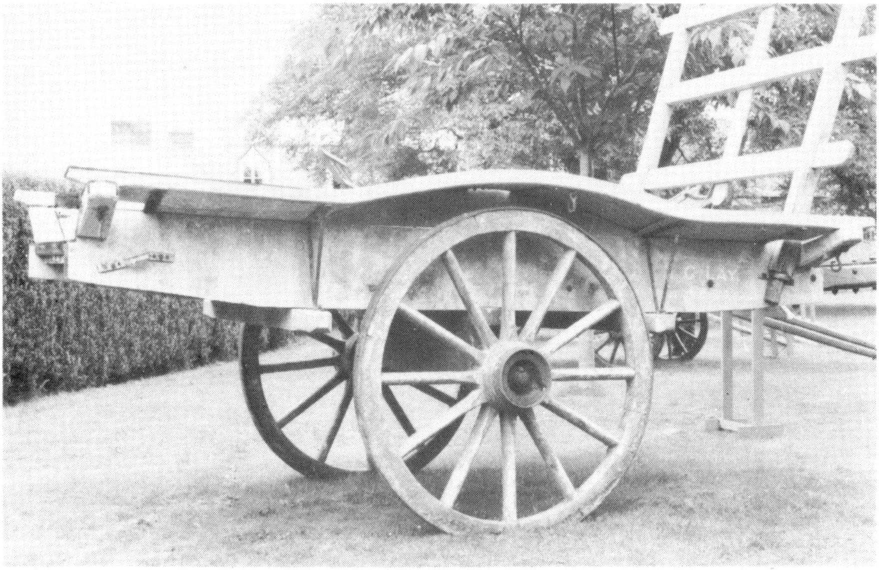

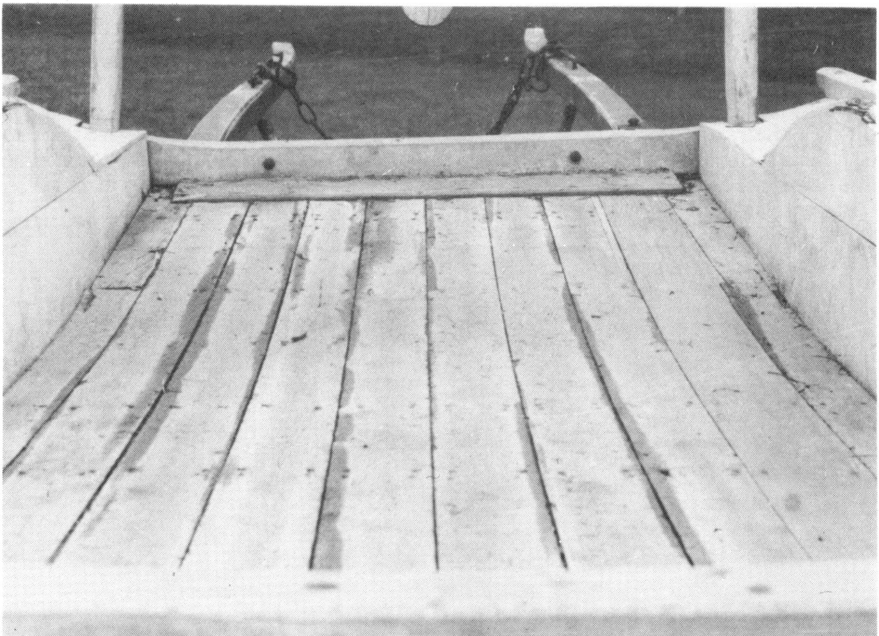

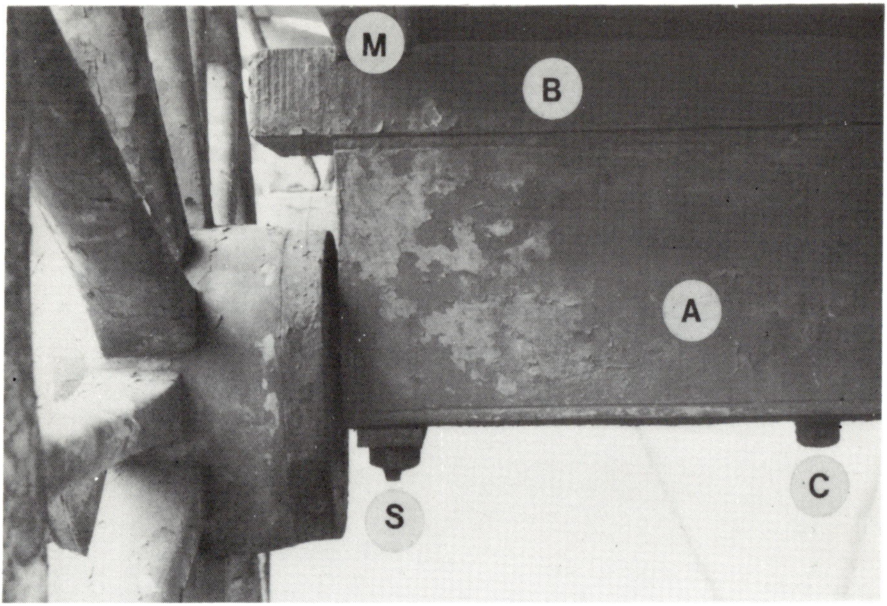

The axle arm of the Cotswold harvest wain is attached to the axletree with two long bolts. Bolt C passes through the axletree A and bolster B. Bolt S also passes through the main side M and the upper side.

Eventually long carts were mass-produced by factory methods. This design, which has low-pitched ladders, is certainly very graceful with its wide flowing outraves. It was produced by the Bristol Wagon and Carriage Works.

This is another long-boarded factory version. It was fitted with a plain but practical harvest frame. Its severe addiction to straight lines shows that it could not be the work of a tradi-tional wheelwright. Nevertheless a vehicle of this type is a good example to start with, for a model with mostly straight members is a much easier construction for a beginner.

This is the earliest long cart to be documented. It was the design adopted by Henry Hannan of Burcot, Oxford-shire, in 1828. From its description we can glean enough information to pro-vide the modeller with its basic dimen-sions. The body was 12 feet (3.66 m) long and 6 feet 6inches (1.98 m) wide (over the outraves). This implies that it had a bed about 4 feet (1.22 m) wide. The sides were not exactly parallel: at the fore end the rails were 6 feet 5 inches (1.96 m) apart; the hooped arch over the wheels was 6 feet 10 inches (2.08 m) across, and the rear rails had a spread of 6 feet 7 inches (2.01 m). From the inside of the main sides the shafts projected 6 feet 8 inches (2.03 m) and their points were 1 foot 10 inches (559 mm) apart. From the sketch shown we can guess that the wheels were about 4 feet 8 inches (1.42 m) in diameter and they were probably at least 3 inches (75 mm) wide.

The cart appears to have been long-boarded and we can infer that it must therefore have had at least two inter-mediate shutters.

A Sussex version of the long cart. This example has a very simply constructed body. The plank sides are held in position by four plain strouters.

In Wales the haycart is known as a gambo. The original vehicle on which this model is based can be seen at the Welsh Folk Museum, St. Fagans, Cardiff.

Another form of the Welsh long cart, with shafts that are also part of the main sides. The body lifts clear of the axle.

A haycart from Jersey. This model was made by Ronald Stiff from the evidence provided by a picture postcard. The medieval-looking sides which protect the wheels can be removed to leave a flat platform like those of the Cornish jack wains.

MORE CART DESIGNS

Livestock was often made to walk to market, but a few interesting carts used for transporting animals have survived. The example above comes from East Anglia and has a low-pitched body, a feature made possible by the special axle. The upper part of the bodywork has an interesting array of turned wooden or plain metal spindles.

Right: A detail of the cranked axle showing how the spring is also attached to the main sides. There are two summers supporting the cross-boarded floor.

Above: A Hertfordshire longshaft timber bob made from David Wray's plan.

Below: Supplying water to outlying buildings and fields presented farmers with constant work in the days before piped water. This water carrier seems to be an amalgamation of a galvanised tank and some rather military-looking wheels. Notice the manner in which the outer ends of the spokes and the felloes are protected. The tank rests upon the cranked axle.

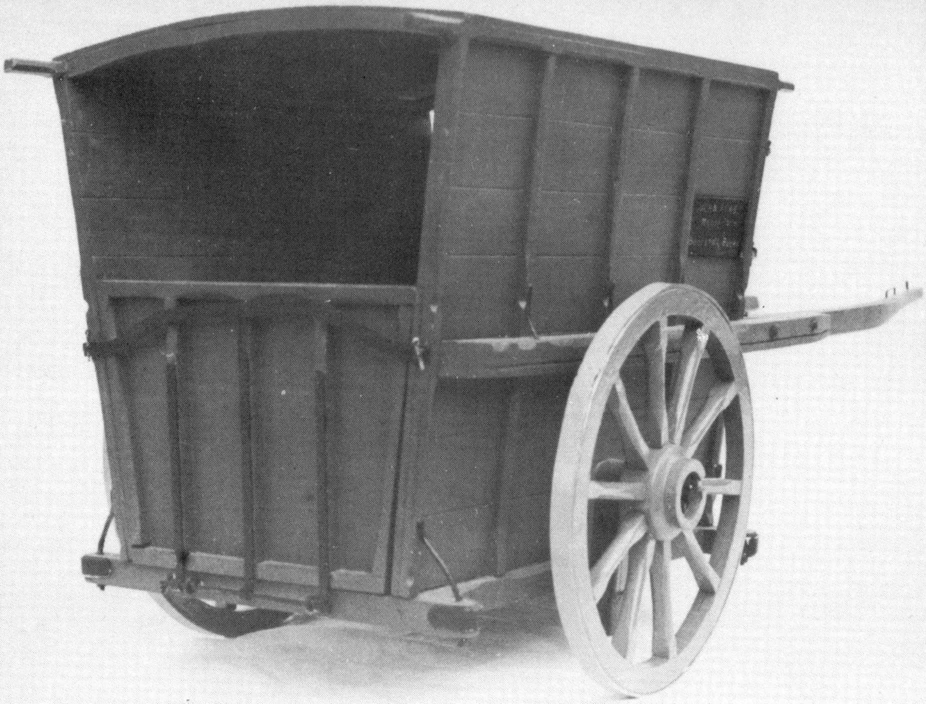

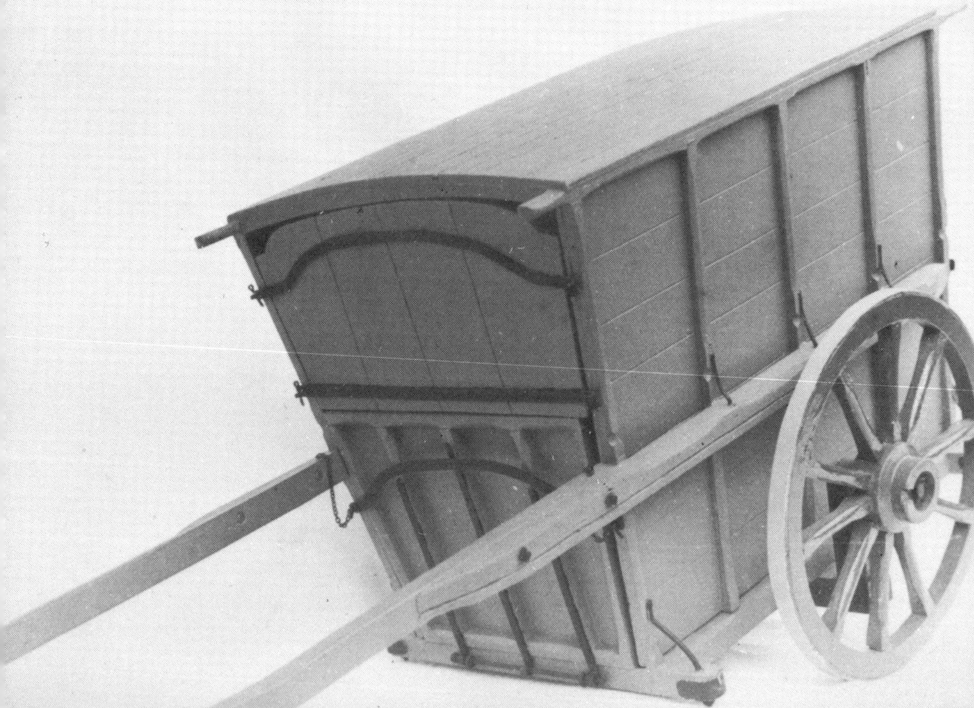

In hilly country the operation of wheeled vehicles presented a number of difficulties. Among the steep combes of Devon it was not uncommon for a considerable amount of soil erosion to take place. The farmers had to spend time each year carrying soil that had been washed down to the bottom of the fields back up to the top. This tedious task would have been impossible with a cart of conventional dimensions and a special vehicle was designed for the purpose.

The gurry butt was designed to be used with a single horse. This example dates from c. 1890. There are no shafts to restrict the horse's movements on the steep and difficult slopes. The gurry butt's small dimensions – about three feet (0.9 m) in height and six feet (1.8 m) in length, limits the burden it can carry. Soil is a dense material to move and on difficult terrain only small loads can be managed. The tipping body is locked with a strap stick.

Opposite: Two views of a bullock cart, from Boarstall, Buckinghamshire. The original body is now used as a garden shed. This design was measured and reproduced in miniature by Ronald Stiff, who has saved it from oblivion. It illustrates the contribution that the modeller can make to historical research.

Above: Stratton's cross-boarded Northampton cart (1854). Notice the unusual tipping device.

Below: This Scottish tilt cart has panel sides and very plain side rails.

Above: A cross-boarded tip cart with panel sides. The space between the toprave and the outrave is boarded in.

Below: A Wiltshire tip cart with a boarded outrave set at an angle to the rather low sides.

BRANDING

The name of the owner almost always appeared on the right side of the cart or wagon just behind the headboard. Some names were written straight on to the bodywork but others were painted on to wooden or metal plates that were nailed in the same position. In some areas very slender name-boards were favoured and these were usually fixed on to the body standards. Lettering used for this plain statement of the name and address was usually unembellished.

On the headboard, however, much greater freedom was enjoyed and there many different forms of lettering could be seen. Not all headboards were lettered. Sometimes the design of the headboard imposed limits on the signwriter. In Glamorgan and the Vale of Berkeley the wagons with a semi-circular panel on the headboard had the inscription confined to this area. Rutland favoured a cartouche, which was usually skilfully filled with its essential message and decorated with scroll designs.

Many regions seem to have used the whole of the headboard for the lettering, which in most cases was executed with great precision. To arrange letters in a symmetrical manner is not easy. A design has to be planned carefully before it is drawn.

Signwriters did not confine themselves to a single alphabet, and it is not unusual to find three distinct styles of lettering appearing on the same headboard. Dates shown need to be interpreted with care. A particular date may not refer to the time of the vehicle's origin but to a major repair or repainting.

The wheelwright would sometimes have his name emblazoned on the tailboad. There were exceptions, like Philips of Flore, Northamptonshire, who placed his name at the bottom of the headboard. When factory-built vehicles came into use metal plates bearing the builders' names were attached to them. Various positions were used for these plates and the older traditions were disregarded.

Lettering can present considerable problems for the modeller. A rather delicate hand is required to execute such fine detail. If a particular model is branded it is probably worthwhile to reproduce the inscription in order to make your model complete.

On cards marked out to the exact shape and size of the area to be covered you can try your hand at lettering on several examples. The lettering is much easier to do before the vehicle is assembled.

Headboards with a plain surface can be given a false headboard after assembly. The branding can then be worked more easily and the finished board attached to the vehicle with glue. Thin card can be used for this purpose. If it is given a coat of varnish it will probably be impossible to tell that it is an addition.

Laying out inscriptions takes time but the effort of trying to master the basic elements of lettering is worthwhile.

Opposite: Various branding styles.
A. When this wagon was repainted someone carefully left the original lettering untouched.
B. A builder's plate of cast iron.
C. A traditional touch with carefully balanced letters.
D. This lettering has a touch of the flamboyant.
E. The tailboard of a Devon wagon where the wheelwright left his mark.
F. Three different styles on one head-board made by a skilled signwriter.
G. An interesting example with letters that slope in several directions.
H. This excellent West Country lettering has an ecclesiastical flavour.
J. A matter-of-fact label from a factory.
K. The tailpiece of a factory wagon with a definite touch of class. The name *Honiton* on the lower scroll has been blocked out. No doubt that was done in 1939.

A

B

C

D

E

F

G

H

J

K

TOOLS, MATERIALS AND DRAWINGS

W. Hobby Ltd
62 Norwood High Street, London SE27 9NW
 Hobby's Annual is a very useful catalogue for the modeller. It contains details of kits, plans, wheels, chains, paint, adhesives, craft knives and saws, lathes, drills and many other modelling needs.
 For fine and precise cutting Hobby's No. 53 Razor Saw Set, which contains two saws, is recommended.

Humbrol Ltd
Marfleet, Hull HU9 5NE
 Humbrol products are well known to most modellers and are stocked in many art and craft shops. The useful Humbrol Major Kit of Craft Tools contains a razor saw, a craft knife with seven blades, a file and tweezers. This item is available from Hobby Ltd above.

Kennion Bros Ltd
2 Railway Place, Hertford SG13 7BT
 Suppliers of brass and mild steel strip and rod; copper and brass tubes; a selection of taps and drills; silver solder; turning tools; rivets and screws; chains and fittings for farm vehicles. Illustrated catalogue (stamped addressed envelope essential).

J. H. Lumsden
Dunroamin, Fron, Garthwell, Montgomery, Powys
 Hand-made scale wheels with elm naves, ash spokes and felloes. Hoop or strake tyres supplied to order. Delivery two or three weeks. Cash with order. Stamped addressed envelope for current charges.

J. Simble & Sons
Queen's Road, Watford, Hertfordshire
 Retailers of every kind of tool imaginable. Illustrated catalogue 40p. Mail order specialists.

Swann-Morton Ltd
Sheffield
 Manufacturers of fine craft knives, which are stocked by most hardware, art and craft shops. The Craft Tool has three different blades – the No. 3 blade is very useful for producing fine chamfers. Unitool and Supatool are appropriate knives for heavier work.

John Thompson
1 Fieldway, Fleet, Hampshire
 Most of the vehicles in his collection of drawings can be seen in rural life museums or in transport museums.
 The plans show all the details clearly. In some cases instructions are included to help the less experienced model maker.
 Ready-made composition and wooden wheels are available. The catalogue of over forty vehicles includes: Wiltshire dung cart, Scotch tip cart, Cotswold harvest cart, Lake District cart, Welsh truckle cart, Forest of Dean wagon, Oxfordshire wagon, hermaphrodite, Hereford wagon, Essex wagon. These plans are available in $\frac{1}{8}$ and $\frac{1}{12}$ scales. Send three first-class postage stamps for illustrated catalogue.

John Vince
c/o Shire Publications Ltd, Cromwell House, Church Street, Princes Risborough, Aylesbury, Buckinghamshire HP17 9AJ
  Measured drawings and explanatory photographs of: Cornish cart, ladder cart, water cart, hoop-raved cart, Aylesbury wagon and others. Send stamped addressed envelope for details.

Barrie Voisey Plans
205 City Road, Fenton, Stoke-on-Trent ST4 3EE
  Suppliers of plans of farm tip cart; Dorset wagon and timber wagon; construction kit for farm tip cart; harness kits; also brass strip and wire and other hard-to-find parts, and ready-made wheels.

K. R. Whiston
New Mills, Stockport SK12 4PT
  Suppliers of a varied selection of materials including: round rod; flat and square aluminium or mild steel; brass and copper tube; BA nuts and bolts; silver solder and flux. Useful bargains offered in each new catalogue, which is a valuable work of reference. Foolscap stamped addressed envelope for details.

David Wray
Little Coldharbour Farmhouse, Berkhamsted Common, Berkhamsted, Hertfordshire
  Illustrated catalogue lists measured drawings of carts and wagons and other horse-drawn vehicles.

BIBLIOGRAPHY

Arnold, James. *Farm Waggons and Carts*. David and Charles, 1977.
Arnold, James. *Farm Waggons of England and Wales*. John Baker, 1969.
Bailey, Jocelyn. *The Village Wheelwright and Carpenter*. Shire Publications, 1975.
Jenkins, J. Geraint. *The English Farm Wagon*. Oakwood Press, 1961.
Sturt, George. *The Wheelwright's Shop*. Cambridge University Press, 1923 and 1963.
Thompson, John. *Making Model Horse Drawn Vehicles*. John Thompson, 1976.
Thompson, John. *Horse Drawn Heavy Goods Vehicles*. John Thompson, 1977.
Thompson, John. *Horse Drawn Trade Vehicles*. John Thompson, 1977.
Vince, John. *Discovering Carts and Wagons*. Shire Publications, 1970.
Vince, John. *Illustrated History of Cart and Wagons*. Spurbooks, 1975.

Model Horse-drawn Vehicle Club
  The club's informative monthly newsletter provides up-to-date details of exhibitions, materials, tools, ideas and information sources of interest to members. It is essential reading for the enthusiast. Send a stamped addressed envelope for deails to: John P. Pearce, 124 Heysham Road, Morecambe, Lancashire.

COLLECTIONS

FULL-SIZE VEHICLES

This list is a selection of museums and other places where carts and wagons may be seen. Intending visitors are advised to check dates and times of opening before making a special journey.

Hereford and Worcester County Museum, Hartlebury Castle, near Kidderminster (telephone: Hartlebury 416).

Horsham Museum, Causeway House, The Causeway, Horsham, West Sussex (telephone: Horsham 4959).

James Countryside Museum, Bicton Gardens, East Budleigh, Devon (telephone: Budleigh Salterton 3881).

Mary Arden's House, Wilmcote, Warwickshire (telephone: Stratford-upon-Avon 3455).

Museum of English Rural Life, University of Reading, Whiteknights Park, Reading (telephone: Reading 85123 ext. 475). Large collection of wagons on display.

Museum of Lincolnshire Life, Burton Road, Lincoln (telephone: Lincoln 28448).

North of England Open Air Museum, Beamish Hall, Beamish, Stanley, Co. Durham (telephone: Stanley 33580 and 33586).

Old Kiln Museum, Reeds Road, Tilford, Farnham, Surrey (telephone: Frensham 2300).

Oxfordshire County Museum, Fletcher's House, Woodstock, Oxfordshire (telephone: Woodstock 811456).

Rutland County Museum, Catmos Street, Oakham, Leicestershire (telephone: Oakham 3654).

Ryedale Folk Museum, Hutton-le-Hole, North Yorkshire (telephone: Lastingham 367).

Welsh Folk Museum, St Fagans, near Cardiff (telephone: Cardiff 561357).

West Yorkshire Folk Museum, Shibden Hall, Shibden Park, Halifax (telephone: Halifax 52246).

White House Museum of Buildings and Country Life, Munslow Aston, Salop.

MODELS

The Ronald Stiff Collection of model farm carts and wagons will, from 1978, be on display at the Oxfordshire County Museum, Fletcher's House, Woodstock, Oxfordshire. This growing collection will eventually include more than fifty vehicles.

Snowshill Manor, Gloucestershire (near Broadway), (National Trust). Among the many attractions at this famous house is the collection of twenty-two model wagons, representing vehicles from the southern counties. They were built by H. R. Whaiting for Mr Charles Wade during the period 1931-8.